BABIES AT THE ZOO
Koala Bear Joeys

Susan H. Gray

Published in the United States of America by
Cherry Lake Publishing
2395 South Huron Parkway, Suite 200, Ann Arbor, MI 48104
www.cherrylakepublishing.com

Content Advisor: Dominique A. Didier, Professor of Biology, Millersville University
Reading Advisor: Marla Conn, MS, Ed, Literacy specialist, Read-Ability, Inc.

Photo credits: ©Anna Levan/Shutterstock.com, front cover; ©Simon Says/Shutterstock.com, 1, 2; ©EQRoy/Shutterstock.com, 4; ©Freder/iStock.com, 6; ©Mari_May/Shutterstock.com, 8; ©covenant/Shutterstock.com, 10; ©Shay Yacobinski/Shutterstock.com, 12; ©Loes Kieboom/Shutterstock.com, 14; ©apple2499/Shutterstock.com, 16; ©Andywak/Shutterstock.com, 18; ©COULANGES/Shutterstock.com, 20

Copyright © 2020 by Cherry Lake Publishing
All rights reserved. No part of this book may be reproduced or utilized in any form or by any means without written permission from the publisher.

Library of Congress Cataloging-in-Publication Data

Names: Gray, Susan Heinrichs, author.
Title: Koala bear joeys / written by Susan H. Gray.
Description: Ann Arbor, Michigan : Cherry Lake Publishing, 2020. | Series: Babies at the zoo | Includes index. | Audience: Grades K-1.
Summary: "Read about the koala bear joeys and how zookeepers take care of them. This level 3 guided reader book includes intriguing facts and adorable photos. Students will develop word recognition and reading skills while learning about how these baby animals learn and grow, what they eat, and how they socialize with each other. Book includes table of contents, glossary, index, author biographies, sidebars, and word list for home and school connection"—Provided by publisher.
Identifiers: LCCN 2019034172 (print) | LCCN 2019034173 (ebook) | ISBN 9781534158955 (hardcover) | ISBN 9781534161252 (paperback) | ISBN 9781534160101 (pdf) | ISBN 9781534162402 (ebook)
Subjects: LCSH: Koala—Infancy—Juvenile literature. | Zoo animals—Infancy—Juvenile literature.
Classification: LCC QL737.M384 G735 2020 (print) | LCC QL737.M384 (ebook) | DDC 599.2/51392—dc23
LC record available at https://lccn.loc.gov/2019034172
LC ebook record available at https://lccn.loc.gov/2019034173

Cherry Lake Publishing would like to acknowledge the work of the Partnership for 21st Century Learning, a Network of Battelle for Kids. Please visit http://www.battelleforkids.org/networks/p21 for more information.

Printed in the United States of America
Corporate Graphics

Table of Contents

5	**Pinkies and Joeys**
13	**Busy Zookeepers**
17	**Leaving Home**
22	Find Out More
22	Glossary
23	Home and School Connection
24	Fast Facts
24	Index

About the Author

Susan H. Gray has a master's degree in zoology. She has written more than 150 reference books for children and especially loves writing about animals. Susan lives in Cabot, Arkansas, with her husband, Michael, and many pets.

Can you name another mammal that has a pouch on its belly?

Pinkies and Joeys

Koalas are **marsupials**. A marsupial mother has a **pouch** on her belly. Her baby lives and grows in it.

Koalas live in Australia. And some live in zoos.

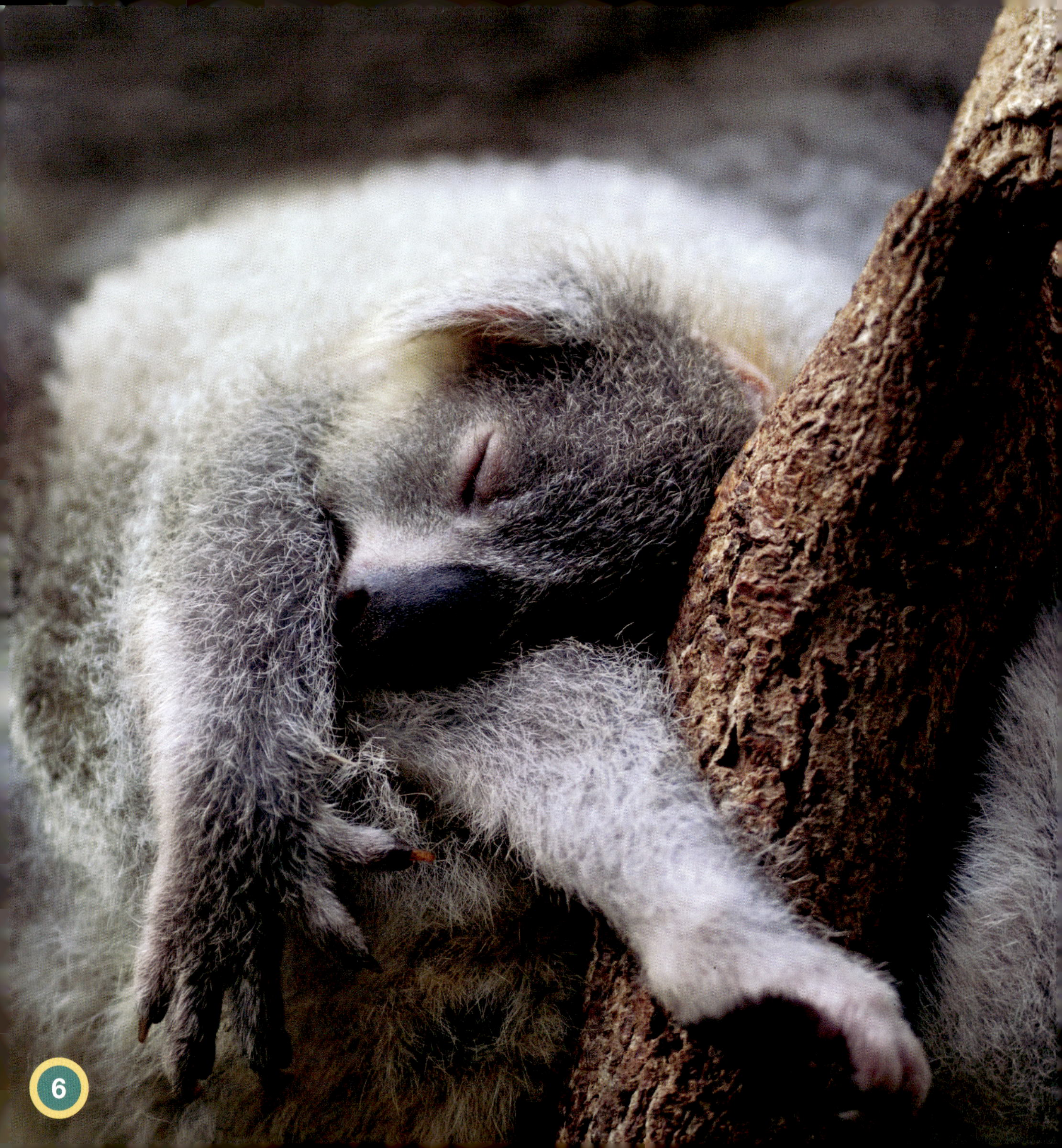

At birth, a baby koala cannot hear and its eyes are closed. But it has strong arms. It can crawl into its mother's pouch to sleep.

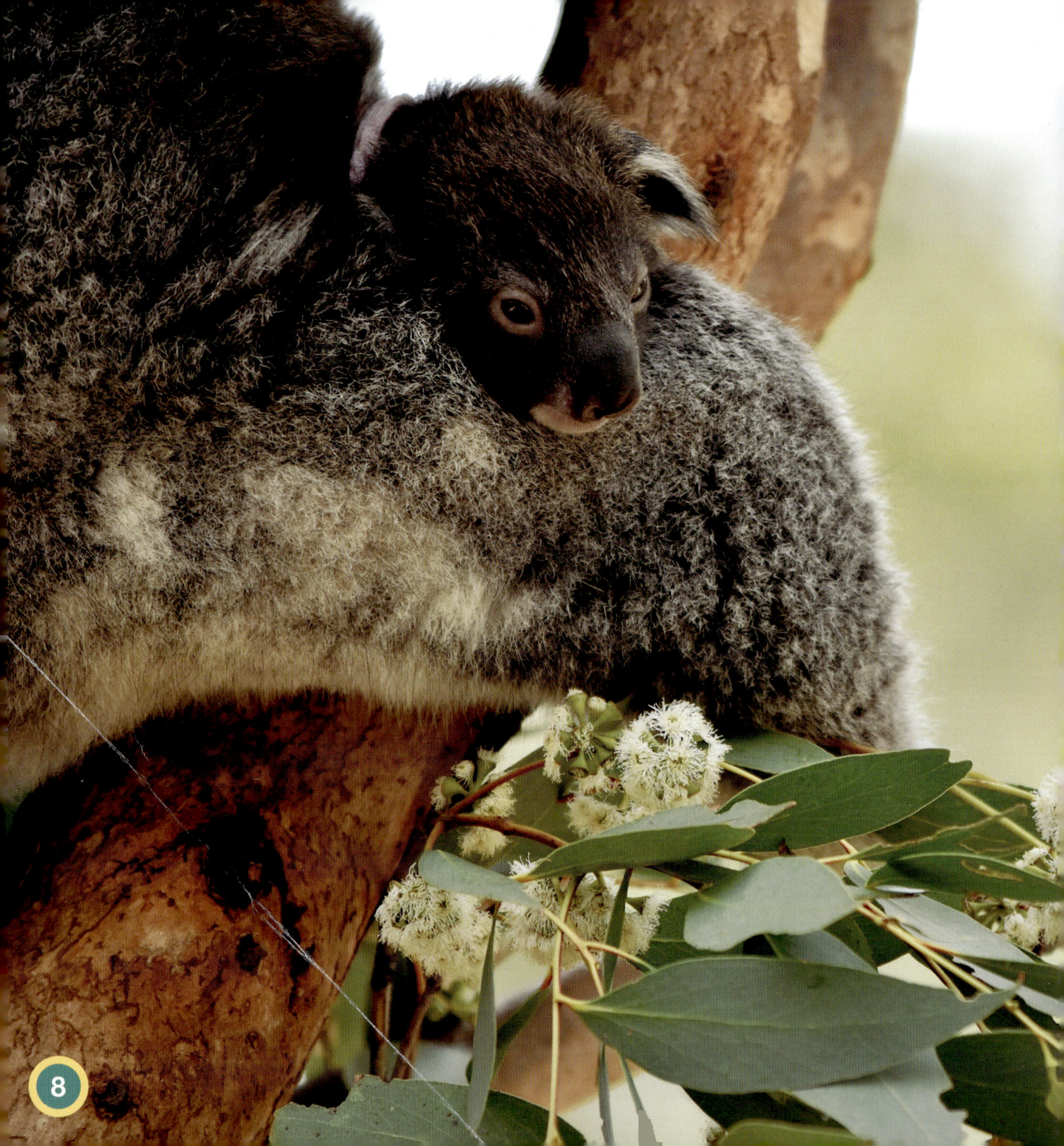

A **newborn** is the size of a jelly bean. Its skin is pink.

All young koalas are called **joeys**. But newborns have a nickname. **Zookeepers** call them **pinkies**.

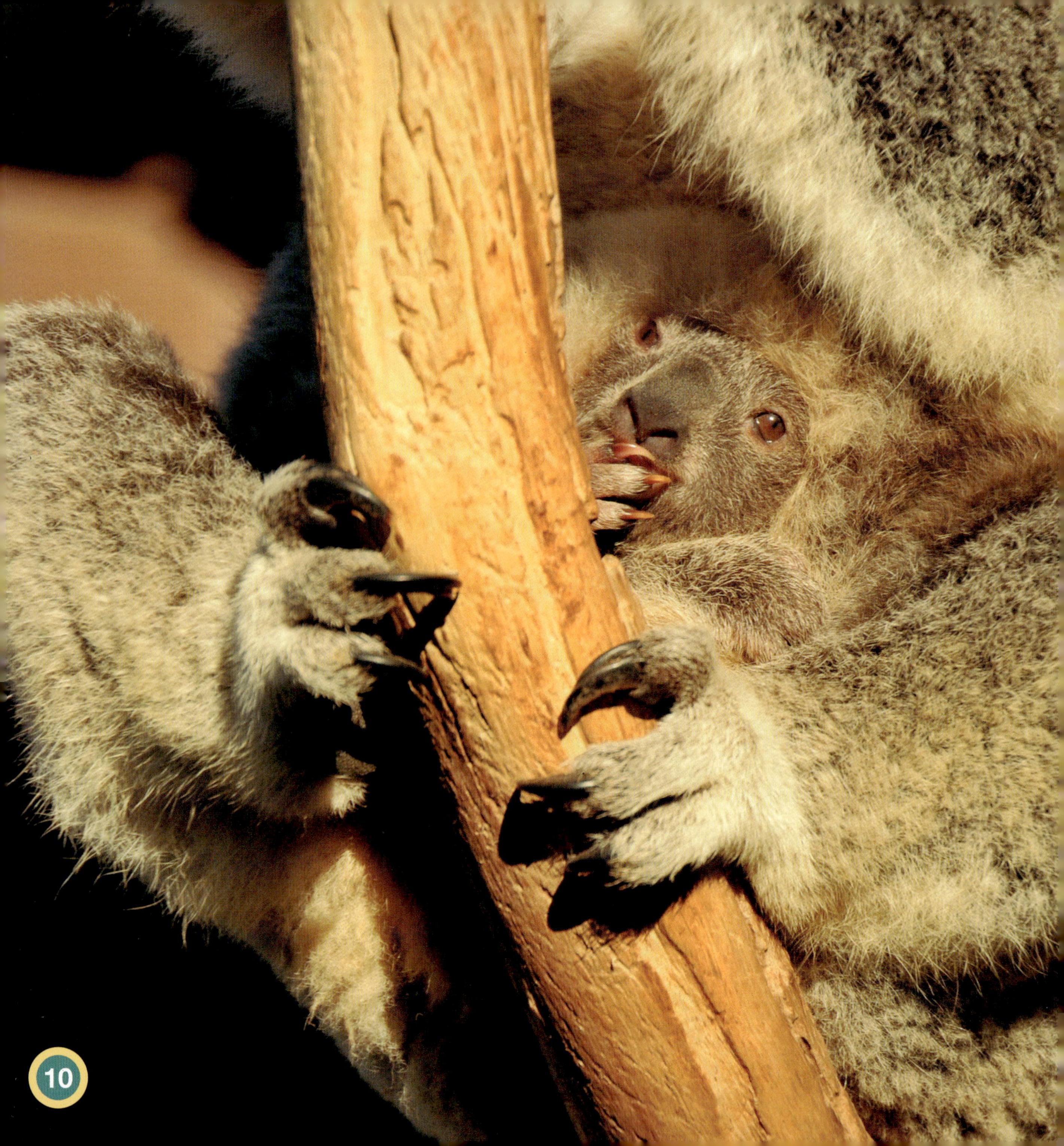

The baby finds milk and safety in the pouch. It stays there for months. Its ears and fur grow. Its eyes open. Little teeth **emerge**.

12

Busy Zookeepers

Koalas live in **eucalyptus** trees. Almost all of their food is eucalyptus leaves. Koalas are picky eaters. The leaves must be fresh. They must smell just right.

What is your favorite thing to climb on the playground?

Zookeepers bring them just the right leaves to eat. They make sure the koalas stay healthy. They even make playgrounds where koalas can climb.

16

Leaving Home

The joey leaves the pouch at 6 or 7 months. But it stays with its mom. It rides around on her belly or back.

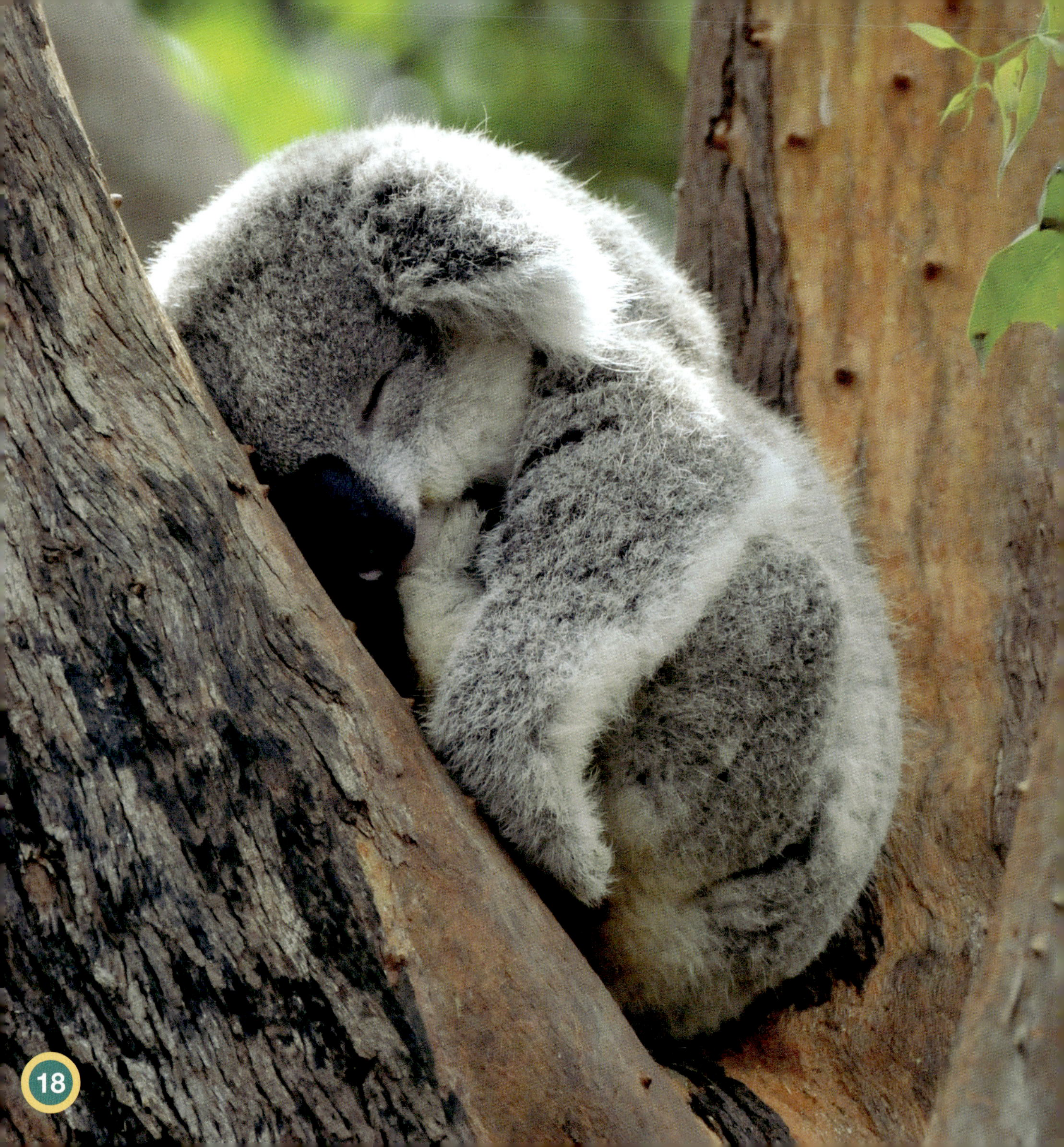

Koalas sleep up to 20 hours a day. They sleep in eucalyptus trees. Joeys often crawl into the pouch to nap.

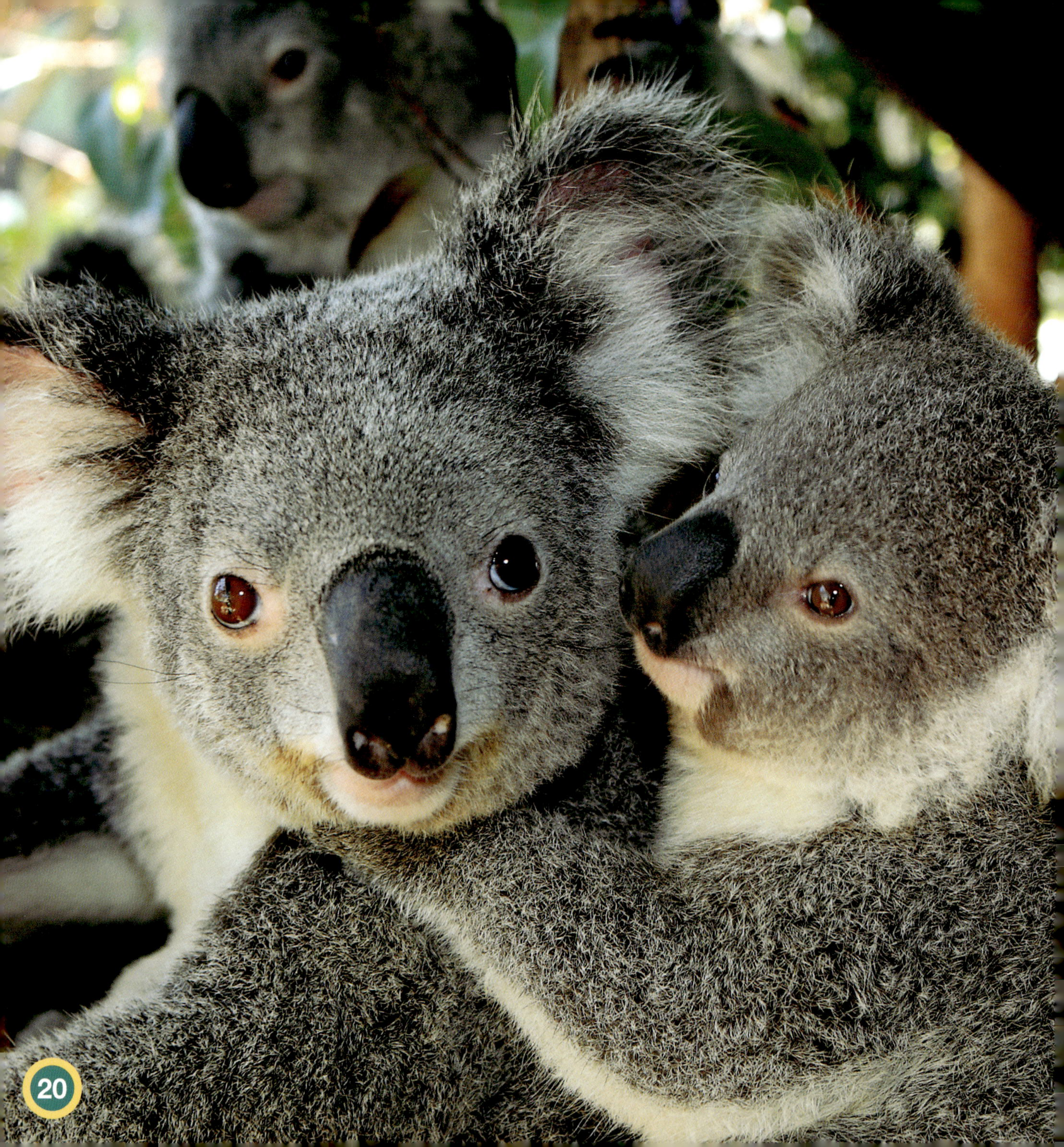

A joey leaves its mom after about a year. It has its own baby a year or 2 later.

And a new, tiny, pink koala enters the world!

Find Out More

BOOK
Marsh, Laura. *Koalas.* Washington, DC: National Geographic, 2014.

WEBSITE
National Geographic Kids—10 Facts About Koalas
https://www.natgeokids.com/uk/discover/animals/general-animals/ten-facts-about-koalas
Find information on baby and adult koalas at this site.

Glossary

emerge (i-MURJ) to come out from a place that is hidden
eucalyptus (yoo-kuh-LIP-tuhs) a tree that normally grows in Australia and has scented leaves
joeys (JOW-eez) baby koalas
marsupials (mar-SOO-pee-uhlz) animals that have a pouch where their young live and grow
newborn (NOO-born) an animal that was born recently
pinkies (PING-keez) newborn koalas
pouch (POWCH) a large pocket
zookeepers (ZOO-kee-purz) people who take care of zoo animals

Home and School Connection

Use this list of words from the book to help your child become a better reader. Word games and writing activities can help beginning readers reinforce literacy skills.

a	called	has	live	own	their
about	can	have	lives	picky	them
after	cannot	healthy	make	pink	there
all	climb	hear	mammal	pinkies	they
almost	closed	her	marsupial	playground	thing
and	crawl	home	marsupials	playgrounds	tiny
another	day	hours	milk	pouch	to
are	ears	in	mom	rides	trees
arms	eaters	into	months	right	up
around	emerge	is	mother	safety	what
at	enters	it	must	size	where
Australia	eucalyptus	its	name	skin	with
baby	even	jelly	nap	sleep	world
back	eyes	joey	new	smell	year
be	favorite	joeys	newborn	some	you
bean	finds	just	newborns	stay	young
belly	food	koala	nickname	stays	your
birth	for	koalas	of	strong	zookeepers
bring	fresh	later	often	sure	zoos
busy	fur	leaves	on	teeth	
but	grow	leaving	open	that	
call	grows	little	or	the	

Fast Facts

Habitat: Eucalyptus forests
Range: Australia
Average Height: 24 to 34 inches (61 to 86 centimeters)
Average Weight: Around 20 pounds (9 kilograms)
Life Span: Around 20 years
Anatomy: A koala has a short, fat tail. Up in a tree, a koala sits on its tail like it's a cushion.
Behavior: Koalas don't need to drink water very often. Most of their water comes from the leaves they eat.

Index

Australia, 5

climbing, 15

ears, 11
eucalyptus trees, 13, 15, 19
eyes, 7, 11

food, 11, 13, 15
fur, 11

joeys, 9
  food, 11
  growing up, 17–21

koala bears
  babies (See joeys)
  food, 13, 15
  where they live, 5, 13
  in zoos, 5, 15

marsupials, 5
milk, 11

mothers, 5, 17, 21

naps, 19
newborns (See joeys)

pinkies, 9
pouch, 5, 7, 11, 17, 19

skin, 9
sleeping, 7, 19

teeth, 11

zoos, 5, 15